YOUR KNOWLEDGE HAS VALUE

AF144009

- We will publish your bachelor's and master's thesis, essays and papers

- Your own eBook and book - sold worldwide in all relevant shops

- Earn money with each sale

Upload your text at www.GRIN.com and publish for free

Abul Hossain, ASW Kurny

Evaluation of Microstructure, Conductivity and Hardness of Al-6Si-0.5Mg-xCu(x=0, 1 & 2) Casting Alloys at Different Ageing Conditions

GRIN Verlag

Bibliografische Information der Deutschen Nationalbibliothek:

Die Deutsche Bibliothek verzeichnet diese Publikation in der Deutschen National-
bibliografie; detaillierte bibliografische Daten sind im Internet über http://dnb.d-
nb.de/ abrufbar.

Dieses Werk sowie alle darin enthaltenen einzelnen Beiträge und Abbildungen
sind urheberrechtlich geschützt. Jede Verwertung, die nicht ausdrücklich vom
Urheberrechtsschutz zugelassen ist, bedarf der vorherigen Zustimmung des Verla-
ges. Das gilt insbesondere für Vervielfältigungen, Bearbeitungen, Übersetzungen,
Mikroverfilmungen, Auswertungen durch Datenbanken und für die Einspeicherung
und Verarbeitung in elektronische Systeme. Alle Rechte, auch die des auszugsweisen
Nachdrucks, der fotomechanischen Wiedergabe (einschließlich Mikrokopie) sowie
der Auswertung durch Datenbanken oder ähnliche Einrichtungen, vorbehalten.

Imprint:

Copyright © 2014 GRIN Verlag GmbH
Druck und Bindung: Books on Demand GmbH, Norderstedt Germany
ISBN: 978-3-656-64509-2

This book at GRIN:

http://www.grin.com/en/e-book/272528/evaluation-of-microstructure-conductivity-
and-hardness-of-al-6si-0-5mg-xcu-x-0

GRIN - Your knowledge has value

Der GRIN Verlag publiziert seit 1998 wissenschaftliche Arbeiten von Studenten, Hochschullehrern und anderen Akademikern als eBook und gedrucktes Buch. Die Verlagswebsite www.grin.com ist die ideale Plattform zur Veröffentlichung von Hausarbeiten, Abschlussarbeiten, wissenschaftlichen Aufsätzen, Dissertationen und Fachbüchern.

Visit us on the internet:

http://www.grin.com/

http://www.facebook.com/grincom

http://www.twitter.com/grin_com

Evaluation of Microstructure, Conductivity and Hardness of Al-6Si-0.5Mg-xCu(x=0, 1 & 2) Casting Alloys at Different Ageing Conditions

Abul Hossain[1] and ASW Kurny[1]

[1]Department of Materials and Metallurgical Engineering,
Bangladesh University of Engineering and Technology,
Dhaka 1000, Bangladesh

ABSTRACT

Factors that influence the microstructure, conductivity and hardness of Al-6Si-0.5Mg-xCu(x=0, 1 & 2) casting alloys were critically analyzed. Experimental improvements of three alloys were studied and the microstructure, conductivity and hardness have been investigated. The addition of Cu to Al-6Si-0.5Mg alloy resulted in the formation of a hard Al2Cu phase and improved the hardness. The Cu-free and Cu-containing Al-6Si-0.5Mg alloys were homogenized (24hr at 500oC), solution treated (2hr at 540oC) and aged at different temperature and time. Formations of different inter-metallic phases were studied by conductivity and hardness measurement in the both Cu free and Cu added Al-6Si-0.5Mg alloys. Inter-metallic phases (hard phases) formed during ageing after solution treatment and they have strong effect on the age hardening and conductivity. An addition of Cu has strong interaction with Al, Si, Mg atoms and vacancies. The increased conductivity in the materials results from the decrease in concentration of the alloying elements in the aluminum matrix due to decomposition of second phase particles and less dislocation density in the formed structure.

Keywords: Al-6Si-0.5Mg alloy, conductivity, age hardening, inter-metallic phases, microstructure.

1. INTRODUCTION

Al-Si-Mg wrought and casting alloys have found extensive use in variety of applications in the automotive, aerospace and defence industries due to their wide range of mechanical and physical properties [1].It is well known that the conductivity is significantly decreased by the addition of alloying elements, whereupon elements in solid solution results in a higher resistance than the same amount of elements forming inter metallic phases. The later typically reduce conductivity proportionally with increasing volume fraction. For Al–Si–Mg and Al–Si–Cu–Mg alloys, a T6 heat treatment consisting of solution treatment, quenching and ageing is often used to increase the strength by precipitating nanometer particles, which provide excellent obstacles for the dislocation movement [2-3]. Among the elements added to Al-Si-Mg alloys for increasing strength and grain-size control, copper has arrested considerable attention. Cu additions reduce the natural aging rate of Al-Mg-Si alloys but generally increase the kinetics of precipitation during artificial aging [4]. In addition to the phases that

1

precipitate in the ternary alloys, the equilibrium precipitate in high Cu Al-Mg-Si-Cu alloys, Q, has been identified as $Al_5Cu_2Mg_8Si_6$ and $Al_4CuMg_5Si_4$ [5-6].

On the other hand, there are several reports regarding the effects of the Cu addition on the age-hardening behavior of Al-Mg-Si alloys. It was confirmed two types of nano clusters in the both Cu-free and Cu-added alloys [7]. Furthermore, some effects such as enhancement of hardness, refinement of precipitates, precipitation sequence and interaction parameters among Cu atoms and solute atoms are investigated in Al-Mg-Si alloys [8-12]. For Al–Si–Mg–Cu alloys, the precipitation behaviors are rather complicated and several phases such as $\beta(Mg_2Si)$, $\theta(CuAl_2)$, S $(CuMgAl_2)$ or Q $(Cu_2Mg_8Si_6Al_5)$ in metastable situations may exist [13–15].

It is well known that the presence of alloying elements or impurities in solid solution increases the electrical resistivity of aluminum alloys. Moreover, the presence of small particles in the structure, such as precipitates or dispersoids, causes significant scattering of conduction electrons and, hence, increases the electrical resistivity of an alloy[16-17].The size and distribution of small particles affect the electrical resistivity of the alloy[18–21].The contribution of precipitates and dispersoids particles could be neglected if the precipitate spacing is greater than the mean free path of the electron, i.e., particle spacings larger than 100 nm in aluminum alloys[17].In the intermediate range of precipitate and dispersoid particle spacings commonly observed in commercial aluminum alloys, the contribution of precipitates to electrical resistivity is less clear. The electrical resistivity analysis can be used as an indirect means of evaluating the volume fraction of dispersoids and the dissolution of constitutive particles during the homogenization of aluminum alloys. There is much research on the effect of stable and metastable precipitates during precipitation hardening on the electrical resistivity of aluminum alloys. All of the researchers have concluded that the presence of elements in the solid solution and small particles possessing coherent inter-faces increases the electrical resistivity or decrease the conductivity. [22–24].

The objective of this research work is to evaluate the effects of Cu and ageing treatment on the evolution of the microstructure, conductivity and hardness of Al-6Si-0.5Mg alloys.

2. EXPERIMENTAL

The Cu-free Al-6Si-0.5Mg alloy and Cu-added Al-6Si-0.5Mg alloy were used in this study. Melting was carried out in a natural gas heating clay-graphite crucible furnace. Several heats were taken for developing the base Al-6Si-0.5Mg alloy. In the process of preparation of the alloys the commercially pure aluminium (99.7% purity) and aluminium-silicon alloy melted in a clay-graphite crucible, Copper in the form of sheet (99.98% purity), was then added by plunging. Magnesium ribbon (99.7% purity) was added into solution duly packed in an Al foil. The final temperature of the melt was always maintained at $900\pm15°C$ with the help of the electronic controller. The melt was degassing with solid hexachloroethane (C_2Cl_6) and homogenized by stirring at 700oC before casting. Casting was done in iron metal moulds preheated to $200°C$. Mould sizes were 15mm x 150mm x 300mm. All the alloys were analysed by wet chemical and spectrochemical methods simultaneously. The chemical compositions of the alloys are given in table1.

The Homogenization, solutionization and ageing of the cast alloys were evaluated using an electrical heating furnace. In this study homogenizing was carried out at 500°C for 24 hours. A set of slices (slice dimension: 20 mm x 20 mm x 8mm) were cut from the three homogenized alloys. The samples were subjected to solution treatment at 540°C for 2hr and rapidly quenched in ice salt water solution. The solutionized samples were kept at room temperature for 1 day prior to artificial ageing. Artificial ageing was conducted in an electrical furnace. For age hardening all the samples were aged at room temperature, 100, 150, 175, 200, 225, 250, 300, 350 400 & 450°C for 1 hour. The other sets of samples were isothermally aged at 225°C for different ageing times ranging from 15 to 360 minutes. The samples were taken out from the furnace and cooled in still air. Subsequently the hardness of each slice was measured. The hardness of all the aged samples was determined on a Rockwell hardness testing machine using the F scale [60 kg load and 1/16 inch steel ball] with a dwell time of 15 s. Hardness measurements were performed on polished slices measuring 20 mm x 20 mm x 8 mm. An average of seven consistent readings was accepted as the representative hardness value of an alloy.

The electrical conductivity of the thermal treated alloys was carried out with an Electric Conductivity Meter, type 979. 20 mm x 20 mm finished surface samples produced by grinding and polishing were prepared for these measurement. These measurements were performed using an eddy-current technique at room temperature (25°C). An Average of five consistent readings was accepted as the representative conductivity value of an alloy. These readings were obtained as a percentage of the International Annealed Copper Standard (%IACS).

Metallographic samples were prepared from selected specimens (after conductivity and hardness measurement) and the microstructures were examined by optical microscopy and scanning electron microscopy with EDX. Samples were prepared by standard metallographic procedures and etched by Keller's reagent (HNO3 – 2.5 cc, HCl – 1.5 cc, HF – 1.0 cc and H2O – 95.0 cc).The washed and dried samples were observed carefully in a microscope at 200X magnification and some selected photomicrographs were taken. SEM of the selected sample was carried out by a Scanning Electron Microscope and the chemical elements of the intermetallic phase were determined.

Table 1. Chemical Composition of the Experimental Alloys

Alloy	Si	Mg	Cu	Ni	Fe	Al
Al-6Si-0.5Mg	5.902	0.461	0.007	0.005	0.146	Bal.
Al-6Si-0.5Mg-Cu	6.105	0.555	1.185	0.029	0.334	Bal.
Al-6Si-0.5Mg-2Cu	5.801	0.497	1.980	0.003	0.300	Bal.

3. RESULTS AND DISCUSSION

Microstructure Al-6Si-0.5Mg, Al-6Si-0.5Mg-Cu and Al-6Si-0.5Mg-2Cu alloys in solution treated condition are presented in Fig. 1, 2 and 3. Analyzing the micrographs of the alloys after solution treatment at 540°C for 2h, it had been found that during solution heat treatment the morphology of primary eutectic Si changes from relatively large needle like structure to the more refined "Chinese script" and spherical in shape particles. α(Al) face-centered-cubic solid solution is the predominant phase (light grey) in the solution treated condition. The silicon-phase which is soluble into aluminium and the other alloying elements form a binary eutectic with α (Al). In the micrographs, the Si eutectic and primary particles are dark grey. The morphology change of the eutectic Si is obvious after solution treatment. The plate like eutectic Si was broken into small particles. The fragmentation process was accelerated by homogenizing (24hr at 500°C) and solutionising (2hr at 540°C) processes.

Al-6Si-0.5Mg-Cu and Al-6Si-0.5Mg-2Cu show eutectic acicular silicon and very few coarse primary silicon particles, embedded in the dendritic aluminum matrix. The Al_2Cu particles are rather coarse, mainly elongated along the grain boundaries and also forming small pockets (Fig. 2 and 3).The structural fineness is seen to increase with increasing copper content. After solution treatment, artificial ageing could enhance the grain refinement (Fig. 4, 5 & 6).The microstructures at peakaged condition (1hour at 225°C) are more refined than the solution treated condition respectively.

Fig.7 shows the characteristics of the intermetallic phases after ageing at 225°C for 1 hour. The complex eutectic mixture experienced fragmentation in the early stage of solution treatment, and a large number of individual particles were formed. EDS analysis revealed that the θ phase dissolved completely after homogenization and solution treatment. During artificial ageing the intermetallic compounds are revealed. The EDS spectra (Fig.8) show the presence of intermetallic compound elements at peak-aged condition. Spot1 (Fig.7) intermetallic compound containing Al, Si & Mg; spot2: Al & Mg; spot3: Al & Si and spot4: Al, Mg & Cu could be identified. So the addition of Mg in Al-Si system indicates the presence of Mg_2Si phases. Cu addition to the Al-Si-Mg system, Al_2Cu, Al_2CuMg, Mg_2Si or other intermetallic phases formed which have a strong effect on hardness, conductivity and grain refinement.

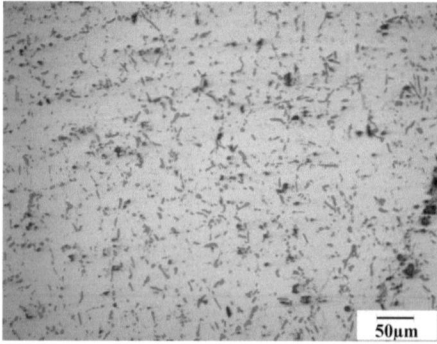

Fig. 1. Microstructure of solution treated Al-6Si-0.5Mg alloy.

4

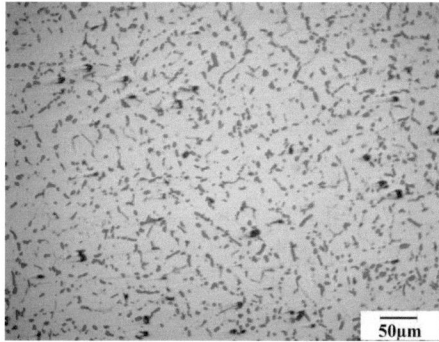

Fig. 2. Microstructure of solution treated Al-6Si-0.5Mg-Cu alloy.

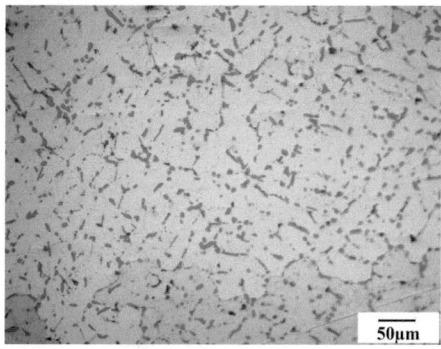

Fig. 3. Microstructure of solution treated Al-6Si-0.5Mg-2Cu alloy.

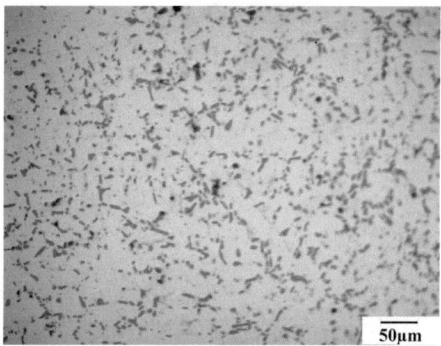

Fig. 4. Microstructure of peakaged (1hr at 225°C) Al-6Si-0.5Mg alloy.

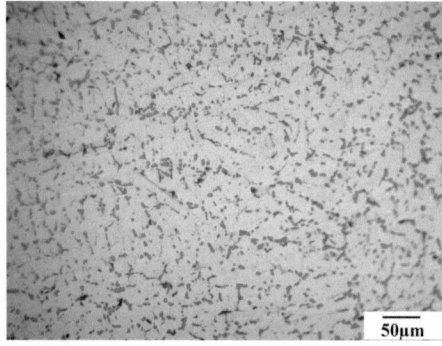

Fig. 5. Microstructure of peakaged (1hr at 225°C) Al-6Si-0.5Mg-Cu alloy.

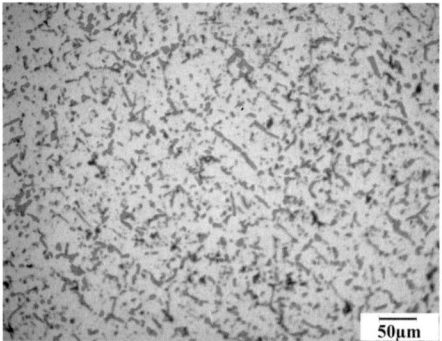

Fig. 6. Microstructure of peakaged (1hr at 225°C) Al-6Si-0.5Mg-2Cu alloy.

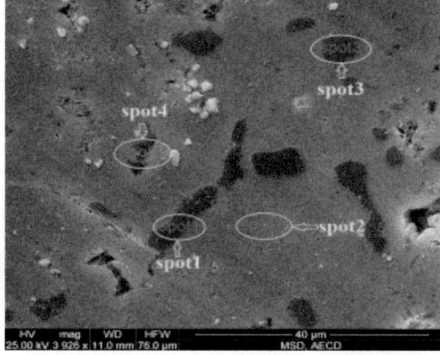

Fig.7. SEM image of Al-6Si-0.5Mg-2Cu alloy at peakaged (1hr at 225°C) condition.

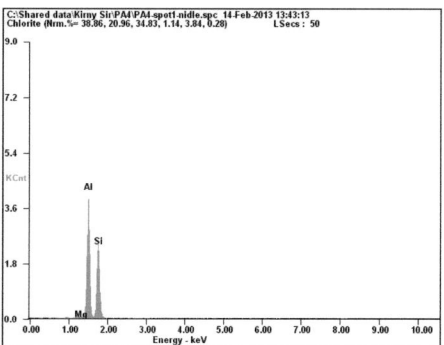

a. EDS spectra of spot1 analysis

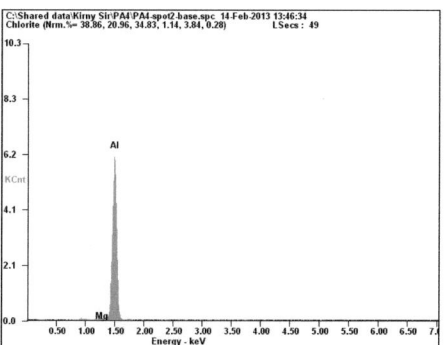

b. EDS spectra of spot2 analysis

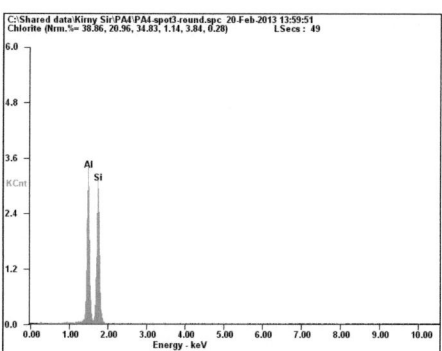

c. EDS spectra of spot3 analysis.

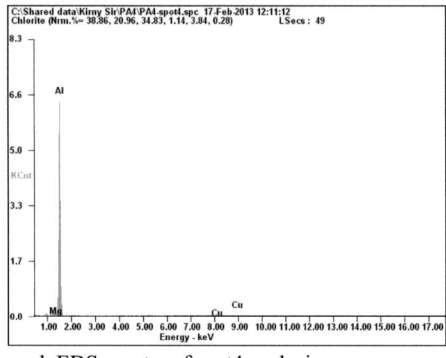

d. EDS spectra of spot4 analysis

Fig.8. EDS spectra of spot analysis (a, b, c & d) of Al-6Si-0.5Mg-2Cu alloy.

Fig.9 and 10 show the change in electrical conductivity and hardness of the quenched and artificially aged Al-6Si-0.5Mg, Al-6Si-0.5Mg-Cu and Al-6Si-0.5Mg-2Cu alloys. Conductivity and hardness were measured immediately after water quenching. The conductivity of the ternary Al-Si-Mg and quaternary Al-Si-Mg-Cu alloys decrease with increasing Cu content. Cu free Al-6Si-0.5Mg alloy (alloy-1) shows the highest conductivity than others Cu containing Al-6Si-0.5Mg-Cu and Al-6Si-0.5Mg-2Cu alloys. So Cu addition reduces the overall conductivity, as can be seen in fig. 9 by suppressions of the curves to lower conductivity values. The effect of grain refinement of the alloys is clearly evident from the conductivity curves, which show a significant difference of conductivity values of the copper added alloy with that of the Cu free alloy. Formation of supersaturated solid solution assures a high precipitation hardening effect upon decomposition of this solid solution with the formation of fine coherent precipitates. The softening of the alloys at higher temperature may be due to particle coarsening effect.

From the age hardening curve (Fig. 10) is shown that hardness in both of the as quenched and peak-aged condition increases strongly with increase of Cu content. The increase of hardness in as quenched conditions with increase of Cu content indicates that the solid-solution strengthening of the alloys is enhanced with increase of Cu content. The maximum hardness is achieved aged at 225°C for 1 hour for the alloys. The decrease in hardness became more pronounced with increasing aging temperature beyond 250°C.The strong increase of maximum hardness during ageing with increase of Cu content is mainly the results of solid-solution strengthening of Cu on a (Al) matrix.

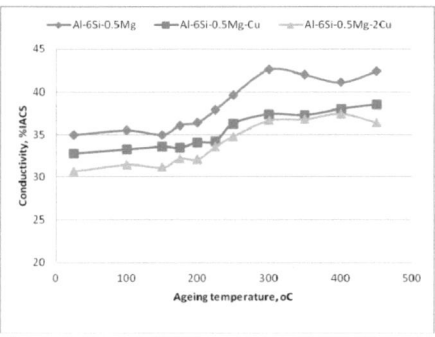

Fig. 9. Variation of electrical conductivity (%IACS) with ageing temperature of the alloys, isochronally aged for 1 hour.

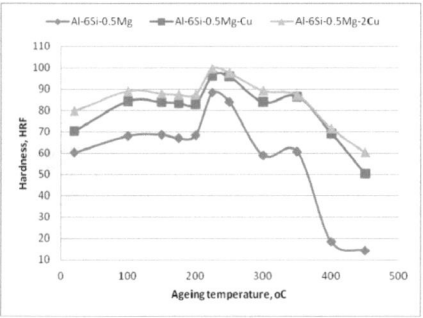

Fig.10. Variation of hardness (HRF) with ageing temperature of the alloys, isochronally aged for 1 hour.

Fig. 11 and 12 show the conductivity and hardness at peakaged condition (1 hour at 225oC).The conductivity and hardness increase linearly with increasing ageing time for isothermally aged at 225oC. Conductivity and hardness change due to form of intermetallic phases while solution treatment and ageing. After 60 minute ageing, the conductivity and hardness of the samples were slightly decreased as the aging time increased and subsequently increased. The change in conductivity and hardness represent the change in the state of solid solution and precipitates in the samples. Therefore, the above results clearly indicate that the Cu solubility is 2 wt% for the Al-6Si-0.5Mg, and it reduces the maximum conductivity compared to others. In Al-Si-Mg solid solution alloys formed precipitation is β(Mg$_2$Si) and Al–Si–Mg–Cu alloys, the precipitation behaviours are rather complicated and several phases such as β(Mg$_2$Si), θ (CuAl$_2$), S (CuMgAl$_2$) or Q (Cu$_2$Mg$_8$Si$_6$Al$_5$) which decrease the conductivity of Al matrix.

9

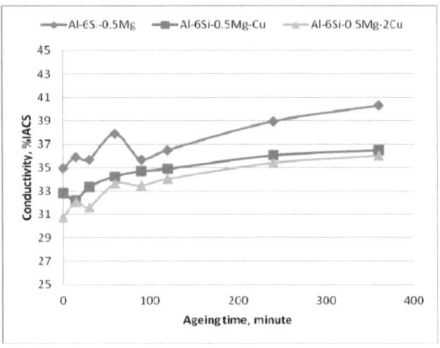

Fig.11. Variation of electrical conductivity (%IACS) with ageing time of the alloys, isothermally aged at 225°C.

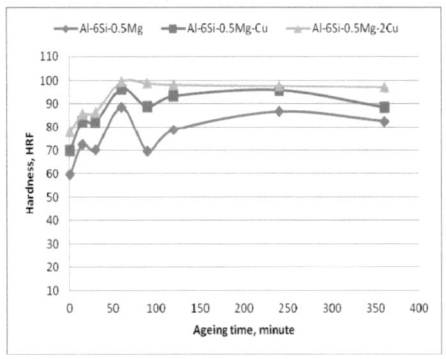

Fig.12. Variation of Hardness (HRF) with ageing time of the alloys, isothermally aged at 225°C.

4. CONCLUSIONS

Addition of Cu to Al-6Si-0.5Mg cast alloy increases the hardness and decreases the conductivity. Maximum hardness was obtained at 225°C for 1 hour ageing for the alloys. The artificial ageing time and temperature strongly effect the conductivity and hardness, because of the eutectic Si, Al_2Cu, Mg_2Si, Al_2CuMg and other intermetallic phases morphology and particles distribution throughout the Al matrix.

REFERENCES

1. Mulazimoglu, M.H., Drew, R.A.L., Gruzelski, J.E., Electrical conductivity of aluminum- rich Al-Si-Mg alloys, *JMSL* , 1989, 8, pp. 297-300
2. Ca´ceres, C.H., Microstructure design and heat treatment selection for casting alloys using the quality index, *J Mater Eng Perform*, 2000, 9, pp.215–21
3. Samuel, P. F.H., Effect of Mg on the ageing behavior of Al–Si– Cu 319 type aluminum casting alloys, *J Mater Sci.,* 1999, 34, pp.4671–97
4. Polmear, I. J., Light Alloys, Metallurgy of the Light Metals, 2nd Ed., Edward Arnold, London, p. 95, (1989).
5. Dumolt, S. D., *Scripta Metall,* 1984,18, p.1347
6. Dutta, I., *J. Mater. Sci. Let.,* 1991, 10, p.323
7. Kim ,J. H. and Sato, T., *J. Nanosci. Nanotechnol,* 2011, 11, pp. 1319– 1322
8. Sato, T., Hirosawa, S., Hirose, K., and Maeguchi, T., *Metall. Mater. Trans. A,* 2003, 31A, pp.2745- 2755
9. Chakrabarti, D. J., and Laughlin, D. E., *Prog. Mater. Sci.,* 2004, 49, pp.389– 410
10. Man, J., Jing, L., and Jie, S. G. J., *Alloy. Compd.,* 2007, 437, pp.146–150
11. Matsuda, K., Uetani, Y., Sato T., and Ikeno, S., *Metall. Mater. Trans. A,* 2001, 47, pp.833–837
12. Miao, M. F., and Laughlin, D. E., *Metall. Mater. Trans A ,* 2000, 31A, pp.361–371
13. Barlow, I.C., Rainforth, W.M., Jones, H., The role of silicon in the formation of the (Al5Cu6Mg2) r phase in Al–Cu–Mg alloys, *J Mater Sci.,* 2000, 35, pp.1413–81
14. Reif, W., Yu, S., Dutkiewicz, J., Ciach, R., Kro´l, J., Pre-ageing of AlSiCuMg alloys in relation to structure and mechanical properties, *Mater Des,* 1997, 18, pp.253–6
15. Mishra, R.K., Smith, G.W., Baxter, W.J., Sachdev, A.K., Franetovic, V., The sequence of precipitation in 339 aluminum castings, *J Mater Sci.,* 2001,36, pp.461–8
16. Zlatanka Martinova, Dimitar Damgalier, Influence of thermo-mechanical treatment on properties of Al-Mg-Si alloy, *Association of Metallurgical Engineers Serbia and Montenegro,* Invited paper AME UDC: 669.715'721'782-156.8=20
17. Sayavur, I., Bakhtiyarov, A., Ruel, Overfelt, and Sorin, G., Teodorescu, Electrical and thermal conductivity measurements on commercial magnesium alloys, *Magnesium Technology 2003* Edited by Howard I. Kaplan TMS (The Minerals, Metals & Materials Society)
18. Lodgaard, L., and Ryum, N., *Mater. Sci. Eng., A,* 2000, 283, pp. 144–52
19. Edwards, J.T., and Hillel, A.J., *Philos. Mag.,* 1977, 35 (5), pp. 1221–29
20. Thakur, A., Raman, R., and Malhorta, S.N., *J. Mater. Process.Technol.,* 2007,194, pp. 184–86
21. Esmaeili, S., Lloyd, D.J., and Poole, W.J., *Mater. Lett.,* 2005, 59(5), pp. 575–77
22. Eivani, A. R., Ahmed, H., Zhou, J., and Duszczyk, J., Correlation between Electrical Resistivity, Particle Dissolution, Precipitation of Dispersoids and Recrystallization Behavior of AA7020 Aluminium Alloy, *Metallurgical and Material transactions A,* 2009, 40, pp.2435-2446
23. Panseri, C., and Federighi, T., *J. Inst. Met.,* 1966,94, pp. 99–107
24. Rossiter, P.L., The Electrical Resistivity of Metals and Alloys, *Cambridge University Press,* Cambridge, United Kingdom, 1987, pp. 1–73